INTELLIGENT COMPUTER COMMUNICATION

Volume 1

FRANK K. APPIAH

ACADEMIC, KWAME NKRUMAH UNIVERSITY.

COLLEGE FACULTY OF COMPUTER
ENGINEERING.

Table of Contents

Illustration Index

Preface

This volume contains the papers presented at ICC2017: Intelligent Computer Communication held on October 9-12, 2017 in Kumasi.

There were 3 submissions. Each submission was reviewed by at least 1, and on the average 1.0, program committee members. The committee decided to accept 3 papers. The program also includes 0 invited talks.

This conference is based on a research and development done in the past 10 years at the Department of Computer Engineering, KNUST. It is part of a doctoral dissertation in my fulfillment of my doctorate degree in 2010-2014.

This conference is based on a book Automatic Information Retrieval System: Intelligent Computer Communication.

This conference was created in Easychair platform.

October 12, 2017   Frank Appiah
Accra.

Dedication

This is dedicated to family and all.

Chapter 1
Essential
Matter

The Essence of the Matter of Intelligent Computer Communication.

FRANK APPIAH , Softgene Technologies Institute Gh. Affiliate: KWAME NKRUMAH UNIVERSITY OF SCIENCE AND TECHNOLOGY, DEPARTMENT OF COMPUTER ENGINEERING, KUMASI, ASHANTI, GHANA.

ABSTRACT.

In this paper, I introduce automatic information retrieval system (AINRS) as an essential matter of intelligent computer communication (ICC). Automatic information retrieval system (AINRS) is a system of information retrieval and communication with an intelligent modem and a communication software by recognition of attention commands in a computer. There is a presentation of the human age in computer and automation especially for human intelligence in section 1.1. Section 1.2 describes the current media of communication that includes texts and pictures used with short messaging service (SMS) and multi-media messaging service (MMS) respectively. The paper shows the aims or objectives that an ICC application helps to achieve in the communication context in section 1.3. In this paper, I discuss scope of work constituent to the system of intelligent computer communication in section 1.4. The paper concludes in section 1.5 with some remarks of computer communication in general.

CCS Concepts:

Intelligent Computer Communication : Application;
Communication : Media

KEYWORDS

ACMICC2017; information retrieval, communication,
computer, intelligent computer communication,
application, automation.

ACM Reference Format:

Prof.Dr. Frank Appiah. 2017. The essence of
matter of intelligent computer
communication. ACM-SIGDOC,XXXX. 12
(October, 2017), 18 pages.

1.1 INTRODUCTION

In this present age, Humans are gradually being replaced by computers in order for tasks and processes to be automated. The fundamental question is : What is the automation of information processing in communication? To have a dynamic intelligent information retrieval system whereby people can retrieve certain kinds of information based on their needs and wants at any convenient time, day or night is one area in which a computer technology will do well to replace a human but no completely. This is the essence of matter of intelligent computer communication. Providing a system in an educational or community facility where communicators or students can make telephone calls or send SMS text messages to access information about their courses as well as courses for a particular semester, time table on specific days and others will reduce the burden students go through in order to access basic information. This system has as its objectives to provide flexible dynamic access to basic information especially in this situation where most communicators or students do not have constant Internet access.

Telephony has come a long way. In this present age of the twentieth century where technology is moving almost at the speed of light, the need for automation has become more pronounced than ever. Services and goods are demanded on the basis of quality rather than quantity. As such automation can now be found in almost every business entity. From the price scanners at supermarkets, level indicators in bottling companies are just a few to mention.

Communication with regard to telephony has evolved into the cellular dimension whereby in a country such as ours it is very uncommon nowadays to find a teenager without a cellular phone. This is not to say that telephones are issues of the past. In actual fact telephones and telephone lines are still being used a lot. The internet for example is one area in which telephone lines are still being used. Bridging the created gap or more to say combining technology from a software aspect to a hardware device such as a telephone line is therefore a best way to combine technologies. Software packages are being developed everyday and as such new programming languages are being developed everyday to make development of these software packages easier. One such language which will be used in this development is the Java programming language. Advancement in Java has made telephony applications easy to deal with either to a first party user such as the provider or to a third party user such the owner of a telephone line. Also two other technologies that can be used for databases are MySQL and XML.

1.2 CURRENT COMMUNICATION (SMS and MMS)

SMS is a text messaging service of most telephone and mobile telephony system. It uses a standardized communication protocols to enable mobile phone devices to exchange short text messages. The protocols allows the sending and receiving of text messages up to 160 alpha-numeric characters to and from the GSM handset. Here, it is a USB GSM modem attached to a computer to create GSM *computerset* or *communicationset*. MMS is a multi-media messaging service that extends the core SMS capability allowing the exchange of short video, image, multi-images and audio. The sending mobile device encodes in a similar MIME fashion and it is then forwarded to a carrier MMS store and forward server called MMS centre. MMS specification considers the flow of peer-to-peer MMS messaging that involves over-the-air transactions but it inefficient for bulk messaging. An executable runtime queue system can help reduce in the overhead transactional bulk-messaging case.

1.3 ICC APPLICATION

Intelligent computer communication ap-plies to the concept of information retrieval system that uses telecommunication service based on current communication services like SMS (for data application) as a text mes-saging component of most telephones, world wide web and mobile technology systems. ICC applies also multi-media messaging ser-vice in it's communication service. ICC appli-cation can be used as the processing soft-ware of distribution list or method to many

number of recipients in multi-media messaging service. The data application of text
messages from mobile devices adds to the
functionality of the ICC system. ICC system
relies on GSM modem to play its major role
of information acquisition and dissemination
from a universal serial bus connection upon
a text message received and a text message
sent respectively. The system indirectly
place services or facility offered on request
by a registered third-party. It is an
uncontrolled intermediary service between
the mobile radio system (provider) and the
mobile handset (user). The intermediary
service is developed to exchange and
process data in a format described by the
modidentifier commands. ICC application
like AINRS is an observatory and analyzer of
data (about 120-160 characters) for
responding to a format request of a mobile
handset. The signaling format of the
application is done by the GSM modem and
stored in the device 's memory. There are
memory issues to address because of the

device's small footprint that is small memory. An ICC application will have to build a fast queuing system most likely on at the computer side to read immediately on request and store in a database or file or data structure for further processing. The scheduling in ICC system most likely will be a round-robin mechanism of design or first-in-first-out FIFO design. This This infrastructure will help accommodate the growing SMS traffic from many request users. The ICC system bases a single modem to a multi-modem device(s) in the communication fabric of the application software by a hub. This is called the *ICC hub or network*. The ICC network implements a large-class of SMS-capable terminal and network. This new element of network works independent of the SMS center. The ICC network attracts the normal charge of the BOSS(Billing and Order Support System) application but an enhancement can be the negotiation of charges to the specific mobile network identification number (mobile phone number) at the SMS center. The ICC application do submit in responds to request by delivering the query format data to a network called specialized short message service center provided by the mobile network provider. Then the mobile application station is created. The ICC application has an over-the-air (OTA) processing dedicated to the network provider. This paper is written to provide the matter of essence to the computer professionals and students :

• For the practicing software engineer, we show how to efficiently develop intelligent computer communication and it's system.
• In the role as a analyst or computer communication engineer, we show you how to effectively implement computer communication systems from the requirement to the design.

• For the student, we provide the instruction necessary for you to begin acquiring several important skills in the developing of intelligent operating system.

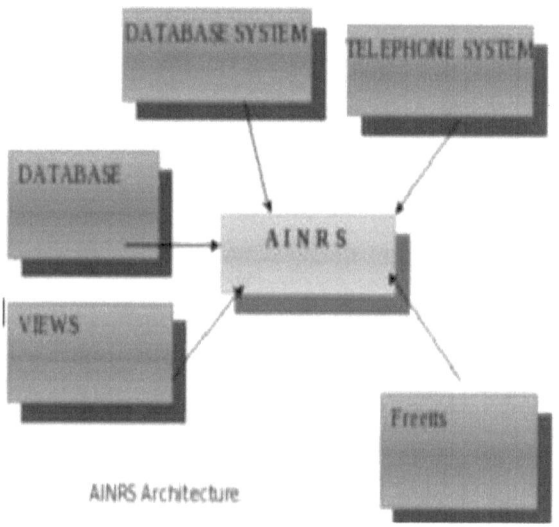

Illustration 1: AINRS Architecture

(c) Frank Appiah. AINRS Architecture. 2008. Drawing.
Author's work of art. Kumasi, Ghana.

To be able to use the software application in different sectors such as banking, laundry services, restaurants and hotels, etc. two components would have to be changed.

These are the database and the set of commands to the system as well as their corresponding configuration class files. These will have to be done in order to be able to generate the correct queries to retrieve the correct data.

1.4 ICC CONSTITUENTS

In this scope, the following constitute work that should be done within the purpose of this paper:

- •Building a custom test database and other related databases for students.

- •Building a telephone system to manage incoming and outgoing DTMF data on the telephone line.

- •Build a mobile system to manage the inflow of requests

and outflow of results.

•Testing of the various units.

•Putting the various units

together and finally testing.

ICC Design will be the determination of processing structures given the desired input and desired output of a system.

Let us now discuss the AINRS architecture to determine the procedural processing of information.

Automatic information retrieval system is made up of 6 architectural components namely, database system (software application) , database models (information structure), views (graphical user interfaces), telephone system (mobile application station, network) and freets (text-to-speech) processing. Each component is responsible for a particular functioning of ICC .

ICC analysis [4] will be the determinant flow of information from the input (mobile

handset component) to the processing components, made up of 6 processing structures. The outputs of each component is to be determined for the program correctness and good functioning.

Let us now look at each program characteristics one-by-one.

MySQL[6] is a database management system. It may be anything from a simple shopping list to a picture gallery or the vast amounts of information in a corporate network. To add, access, and process data stored in a computer database, you need a database management system such as MySQL. It will provide the output / input correctness of the program. The information in a database will be sent to a mobile handset as it appears in table or mobile handset will request a save of information to a table in a database. A database is a structured collection of data. [1] MySQL is a relational database manage-ment system. A relational database stores data in separate tables rather than putting all the data in one big storeroom. This adds speed and flexibility.

The tables are linked by defined relations making it possible to combine data from several tables on request. MySQL is a client/server system that consists of a multi-threaded SQL server that supports different back ends, several different client programs and libraries, administrative tools, and several programming interfaces.

Now let me discuss the telephony side of the ICC system.

The Java Telephony API [5] (JTAPI) is a portable, object-oriented application programming interface for Java-based computer-telephony applications. Similar APIs for other platforms are the Telephony API (TAPI) for the Microsoft Windows platform and the TS-API for the Novell Netware platform.

SMSLib [7] is a Java compatible library that enables the sending and receiving of SMS messages via a GSM phone or GSM modem.

A typical deployment consists of:

1. Computer with free COM port
2. Mobile phone, which can be connected to COM port

3. Cable to connect mobile phone to the COM port.

The computer and the mobile/GSM modem forms the Mobile Terminal (MT). The Mobile Adaptor is the cable/USB connecting the mobile. The library has support for several phones simultaneously. The concept of gateway is been introduced. This is an interface to a device or service that can send and/or receive SMS messages. A gateway could be a GSM modem or a supported bulk SMS provider. SMSLib can handle multiple gateways at the same time. Let us conclude now.

1.5 CONCLUSION REMARKS

The ICC core is made up three main parts namely: the database , telephony and executable/ runtime software. The database is implemented with MySQL and persistence library of Java (database model). The telephony part of the ICC core is implemented with JTAPI and SMSLib of Java. The runtime software is made up of executable code of Java and it does the real processing of information.

REFERENCES

[1] Database System Concepts by Henry F. Korth and Abraham Silberschatz.

[2] Digital Data Communications by Jack Quinn.

[3] Modern Digital and Analog Communication Systems, B. P. Lathi.

[4] Computer systems by J. Stanley Warford, 2005, Jones and Barlett Publishers, Inc.

[5] Essential JTAPI, JAVA™ TELEPHONY API by Spencer Roberts [6] http://www.mysql.com/ , further information on Mysql and Downloads (Community Server Version).

[7] http://smslib.org. For SMS API

Chapter 2
Technology
<u>Survey</u>

A Survey Review of Technology Literature in ICC.

FRANK APPIAH , *Softgene Technologies Institute Gh.Affiliate: KWAME NKRUMAH UNIVERSITY OF SCIENCE AND TECHNOLOGY, DEPARTMENT OF COMPUTER ENGINEERING, KUMASI, ASHANTI, GHANA.*

ABSTRACT.

This survey paper starts with a short literature review of Java technology, MySQL database application, XML, telephony system, SMS technology and modem hardware technology in section 2.1. The question of why Java is asked in the review, then the other question of Why XML is also asked/answered and the third question of why MySQL is also answered. Section 2.2 of the paper gives a brief description of data communication and principles of digital data communication. Section 2.3 describes the cellular telephone system or mobile radio system as brief survey. Section 2.4 is about the conclusion remarks of the technology literature of ICC and furthermore detail workings of the character-command computation is in section 2.5.

CCS Concepts:

Intelligent Computer Communication : communication technology

KEYWORDS

ACMICC2017, intelligent; computer; communication; technology; database; telephony; system; messaging; language.

ACM Reference Format:

Frank Appiah. 2017. A survey review of technology literature in ICC. ACM-SIGDOC,XXXX. 12 (October, 2017), 37 pages.

2.1 LITERRATURE REVIEW

This paper will look at the survey review of some key technologies in building an intelligent computer communication systems in a much literature. "Why Java" is first in review, then "Why MySQL" is second in survey and more in this section.

2.1.1 Why Java?

The Java Programming language has come along a long as its predecessor and rival counterpart C++. Next to Bell Laboratories C++, it is the next most popular and well supported programming language available today. It's support spans from its numerous thousands of programmers and its active forums. It provides support interfacing with hardware through the ports on the computer. It is a language that also has a lot of support for client server interactions and distribution of resources in such applications. Java being built on C++ robustness and clean syntax implies that it uses most of the best features from that language.

2.1.2 Why MYSQL?

Relational database engines are nowadays very popular since many firms and organizations are now keeping their data using relational concepts. Among the most famous engines include Oracle, Microsoft's SQLSERVER, etc. The choice for using MySQL[3] is simply to reduce cost in this development. MySQL is free and also an open source project. This allows the third party developer to make modifications as well as additions to the engine during his development.

2.1.3   Why XML?

Extensible mark-up language, popularly known as XML is a mark-up language that allows developers and its user to focus definitions on relationships rather than on presentation. Using XML, one does not have to focus on how the data will appear in for example a web page. XML is being used in this project because of its lightness. Compared to traditional database engines as discussed above XML being used for storing data just for retrieval purposes makes it lighter to the operating system and memory management processes.

2.1.4   Telephony System

Telephony encompasses the use of equipment to provide voice communication over distance by connecting telephones to each other. Telephony has come a long way from when telephones were initially connected in

pairs so that individuals had many telephones to places they wanted to reach. This
then evolved to the exchange system where
all telephones were connected to a
particular unit making calls to be redirected.
In modern times such these, analog
telephony has gradually been replaced by
digital telephony both in telephone services
and systems. This has cut down cost and
improved the quality of voice services today.
Also because of the digitization, a lot of
technologies have been developed around
telephone equipment. Telephone lines are
now being used to transfer data from one
computer unit to the other. An example of a
technology that uses telephone lines is the
Internet.

The telephone network is made up of all of
the world's telephone companies called **telcos** for short. Customers of the telcos are
called **subscribers**. There are two types of
services:

1.   Access to the worldwide *dir-
ect distance dial* (DDD) network,

which is the type of telephone service.

2.   *Private line service*, which
provides dedicated, semi-perman-
ent telephone connections between
fixed points.

The three main components of the DDD network are station equipment, switching equipment and transmission equipment. Station
equipment includes all parts of the telephone network that are located on the subscriber's premises such as the telephone set
itself, private branch exchange and the
wiring within the subscriber's home or business. Switching equipment is located in telco
offices and makes the connection between
the station equipment of two or more subscribers. Transmission equipment consists of
the telephone circuits that carry information
from one location to another. There are two
types of telephone circuits :

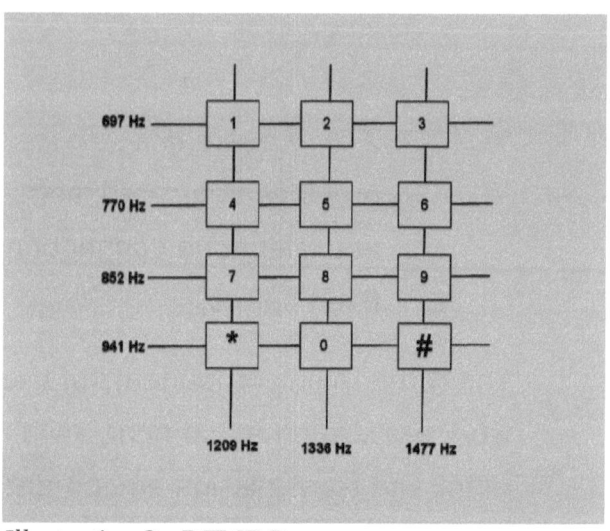

Illustration 2: DTMF Buttons

**(c) Frank Appiah. DTMF Buttons. 2008. Drawing.
Author's work of art. Accra, Ghana.**

local loops and trunks. Local loops connect the subscribers' station equipment to the switching equipment in the telco local office. Trunks carry communications from one item of switching equipment to another or from one local office to another. Trunks are circuits that connect large subscriber's private branch exchange to the local office. A tele-

1   Quinn, Jack.DTMF Buttons. 2017. Digital Data Communication
    Book.

phone is also called telset. The main components of a telset are the ringer, which sounds to alert the subscriber when there is an incoming call; the hook switch, which is open when the telephone handset is "on the hook" and which closes when the handset is lifted; an ear piece called a receiver; and a microphone called a transmitter. In modern telset, pulse dialling has been replaced by dual-tone multi-frequency signalling (DTMF), more commonly known as **Touch Tone**. DTMF illustrated above uses seven audio oscillators, one for each of the four rows and one for each of the three columns of buttons on the telset. Each time a number is dialled, two oscillators are activated simultaneously, one for the row in which the button is located and one for the column in which it is located. If 3 is dialled, both 697 Hz and 1477 Hz oscillators sound and the two frequencies that they generate are sent to the telephone line. DTMF tones can be sent more rapidly than dial pulses.

2.1.5 Short Messaging Service

Short Messaging Service (SMS) is a technology that enables short messages of about
160 characters to be sent from one mobile
phone to the other. It can be used to send
regular text as well advanced content such
as operator logos, ringing tones, phone con-
figurations etc.

SMS Appeared in Europe in 1991. The Global
System for mobile communications included
SMS from the outset. SMS is built on digital
wireless standards such as code division
multiple access (CDMA), time division multiple access (TDMA) and GSM (Global system
for mobile communication). SMS provides a
mechanism for transmitting data to and from
wireless devices. This makes use of a short
messaging servicing centre (SMSC). The
SMSC acts as a store and forward system for
short messages. In contrast to other existing
text messaging services such as alphanu-
meric-paging, the service elements are de-
signed to provide guaranteed delivery of
text messages to the destination. A very dis-

tinguishing aspect of SMS is that a mobile phone or handset is able to deliver or receive messages irrespective of whether a voice or data call is in session or not. SMS is characterized by out of band packet delivery and low bandwidth transfer which results in efficient means of transmitting short burst of data.

2.1.6   Modem

Modem is an acronym that stands for modulator demodulator. A modem is a simple device which is capable of converting analogue signals to digital ones and digital signals to analog. Modem is also called *data set*. Modems used to be a separate hardware third party users that had to be purchased separately. However, today's computer manufacturers have made it one of the primary hardware components found on any

computer. It is very rare to come across a
computer without a modem.

The modem performs five basic functions:

1.   It accepts transmitted serial
digital data that the DTE sends, often by means of an RS-232 transmit
data (TD) line.

2.   It modulates the data onto an
analog carrier and transmits the
data over an analog communications medium such as a radio channel or a telephone line.

3.   It demodulates the analog
carrier that it receives from the
analog communication medium to
recover the received digital data.

 *Illustration 3: USB GSM modem*	 *Illustration 4: External Fax /Voice /Data modem*

(c) Frank Appiah. Lightwave Modem. 2008. Photography. Author's work. Kumasi, Ghana.

4.   It passes the demodulated serial data to the DTE, often by means of an RS-232 receive data (RD) line.

5.   In synchronous systems, the DCE supplies a clock signal to the DTE.

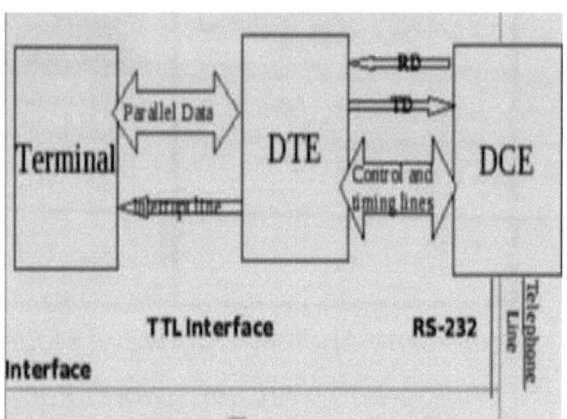

Illustration 5: Physical Form of Modem

Source: Quinn, Jack.DTMF Buttons. 2017. Digital Data Communication Book. Drawing. Author's art.

Modems are built in three physical forms. These are mainly :

> • Built-in modem: uses the same logic levels as the computer itself. The advantages of built-in modems are that they are inexpensive and are an integral part of the computer. The disadvantage of built-in modems is that it is difficult to upgrade then when newer and faster models are introduced. Built-in modems connect directly to the

computer's internal parallel bus.

• **Plugin modem** : It includes the data terminal equipment(DTE) and they use the same logic as the computer. Plug-in modems connect directly to the computer's internal parallel bus. They are therefore in intimate communication with the computer.

• **Stand-alone modem** : It connects to the terminal by means of an RS-232 interface. Stand-alone modems do not include the DTE. The DTE is located within the terminal. Stand-alone modems have their own internal power supplies and some derive their power from the RS- 232 interface signals supplied by the terminal.

Simplex, Half-Duplex and Full-Duplex Modems

Modems can be simplex, half-duplex or full-duplex. Simplex modems are used in applications such weather wires and news service wires. Because simplex communication is one direction only, simplex modems can use the full bandwidth of the telephone circuit.

A disadvantage of half-duplex communication is that each time the direction of communication is reversed, the telephone circuit must be "turned around". The modem that was transmitting is switched to the receive mode, and the modem that was receiving is switched to the transmit mode. Turning a circuit around can take several hundred milliseconds, a waste of valuable communication time. The second disadvantage of half-duplex communication is that the receiving terminal cannot provide immediate feedback in case of errors. Some half-duplex modems have a low-speed channel that is too slow

for data communication. It is adequate to al-
low the receiving terminal to signal the
transmitter that a block of data was either
received correctly or was received in error.

Full-Duplex modems transmit and receive
at the same time. When a full-duplex modem communicates over a four-wire leased
telephone circuit, it can use the full bandwidth of the communication channel. For
full-duplex communications over two-wire
telephone circuit, technology once required
modems to use one carrier frequency to
transmit and a different carrier frequency to
receive.

The illustration below shows how the modem divide the telephone line into two communications channels, which are labeled
east-to-west and west-to-east. Each channel
uses half the bandwidth of the telephone
line.

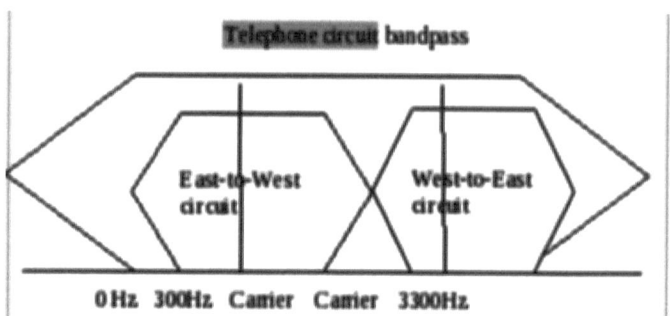

Illustration 6: Telephone circuit bandpass

Source: Quinn, Jack.DTMF Buttons. 2017. Digital Data Communication Book. Drawing. Author's art.

The maximum possible communications speed is proportional to the bandwidth of the communications circuit. The next is a digital modem built with a digital signal processor that performs the functions of modulation , demodulation and echo cancellation. Digital data in serial format are input to the DSP, and the DSP generates a digital representation of the modulated carrier. The digital-to-analog converter (D/A) converts the digital output of the DSP to an analog signal, which passes through the hybrid circuit and onto the two-wire telephone line.

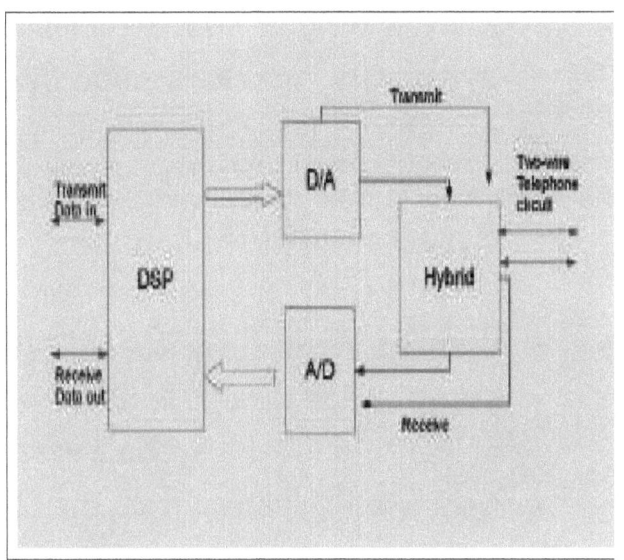

Illustration 7: Hybrid Circuit

**Source: B. P. Lathi. 2017 Modern Digital and
Analog Communication Systems Book. Drawing.
Author's art.**

A hybrid circuit is a circuit that connects a
four-wire to a two-wire line. In the figure, the
hybrid passes the transmitted signal to the
two-wire telephone line and blocks it from

entering the modem's receive section and it routes the signal from the telephone line to the modem's receive section. The receive carrier from the two-wire telephone circuit passes through the hybrid to the analog-to-digital converter (A/D), which converts it to a digital format. The DSP demodulates the digitized received signal and outputs the received digital data in serial format. In cases of leak through the hybrid and reflection from the telephone circuit, the DSP detects and remove any echoes of the transmitted signal that it finds in the received signal.

A modem modulates serial digital data onto a sine-wave carrier. Modulation occurs when the data vary some characteristic of the sine wave. The three sine-wave characteristics that can be varied are amplitude, frequency and phase. Frequency modulation in data communication is frequency-shift keying (FSK) and phase modulation in data communication is also called phase-shift keying (PSK). Modern high-speed modems use a combination of AM and PSK.

One method used to increase the apparent speed of a modem is data compression. Simply, data compression is a method of using fewer bits to represent the same information. Synchronous modems need a clock signal of the proper frequency and phase to modulate and demodulate. FSK modems are asynchronous. They do not require a clock signal to frequency modulate the data onto the carrier. They do not require a clock signal to demodulate the received data.

Modem Standards

Modem standards can be divided into three types:

> 1.   Standard methods for controlling the operations of a modem.

> 2.   Standard methods for compressing data and checking it for transmission errors.

3.   Standard methods for modu-
lating data onto a carrier at the
sending modem and demodulating
it at the receiving end.

Bell system of telephone companies and
the Consultative Committee for International
Telegraph and Telephone (CCITT) are the two
organizations that has set modem standards. The controlling of a modem is based
on computer software that recognizes characters sent to it as commands or a system of
common commands called At command set
which is also known as Hayes command set.
AT command is an attention command. This
behavior makes the modem, an intelligent
modem capable of dialing a telephone number, recognizing a busy signal, answering an
incoming calls, correct errors and performing
many other functions.

When the modem is not communicating
with another modem, it is in the command
mode and it switched to an on-line mode
when it establishes a communication with a

remote modem. The modem treats characters from the data terminal equipment by in-
cluding AT characters, as data prefix.

Some common commands in the AT command set

Character(s)	Command
AT	Attention.
A	Answer an incoming call.
DT	Dial using DTMF tones.
EO	Do not echo transmitted data to terminal screen.
E1	Echo transmitted data to terminal screen.
FO	Half-Duplex communications.
F1	Full-duplex communications.
H	Go on hook (hang up).
DP	Dial using pulse dialing.

All of the remaining AT command set are in [2].

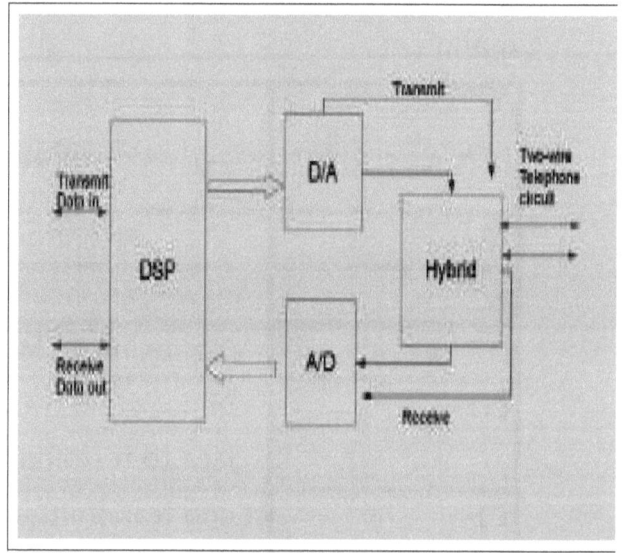

Illustration 8: Hybrid Circuit

2.2 DATA COMMUNICATION

In the field of data communication the word data means information that is stored in digital form. The contents of a computer memory is a data. A single unit of data is datum and data is the plural of datum. Data is a denote of a certain type of information. The transfer of information that is in digital form before it enters the communication system is called data communication.

In the field of data communication the word data means information that is stored in digital form. The contents of a computer memory is a data. A single unit of data is datum and data is the plural of datum. Data is a denote of a certain type of information. The transfer of information that is in digital form before it enters the communication system is called data communication.

In 1948, Claude Shannon developed a mathematical equation to calculate the maximum rate at which data can be communicated over a given communications channel. The equation is known as **Shannon' Law**. It states that :

$$C(b/s) = BW \log_2 (1 + S/N), \text{ where } C = \text{in-}$$

formation-carrying capacity, BW= bandwidth, S= signal power and N= noise power. B/s is bits per second.

The formula shows that the information-carrying capacity of a communication link is proportional to its bandwidth.

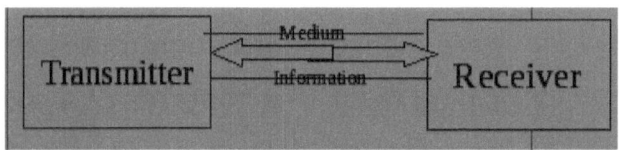

Illustration 9: Transmitter-Reciever

Source: B. P. Lathi. 2017. Moden Digital and Analog Communication Systems. Drawing.Author's art.

Principles of Digital Data transmission The digital data transmission considers the use of binary case where data consists of only two symbols or values: **1** and **0** at the trans-mitter and receiver ends. A digital communi-cation system is made up of :

- **Source** : The input to a digital system is in the form of a sequence of digits. The input and output are data sets, computers, a digitized voice signals (PCM or DM), telemetry equipments etc.

* **Multiplexer** : The digital multiplexer using the process of interleaving combines several sources to utilize the larger data rate of the channels than the individual sources. Thus, a channel is time-shared by several messages simultaneously.

* **Regenerative Repeater** : Regenerative repeaters are used at regularly spaced intervals along a digital transmission line to detect the incoming digital signal and regenerate new clean pulses for further transmission along the line. The periodic way of working eliminates and thereby combats, the accumulation of noises and signal distortion along the transmission path.

- *Line Coder :* The output of a multiplexer is coded into electrical pulses or waveforms for the purpose of transmission over the channel. The process is called line coding or transmission coding. There are five types of line codes:

- *On-Off* (Return-to-Zero: **RZ**): A 1 is transmitted by a pulse p(t) and a 0 is transmitted by a no pulse (zero signal). This code has the chance to go to zero before the next pulse begins, that is On-off (RZ) scheme.

- *Polar* (Return-to-Zero) : A 1 is transmitted by a pulse p(t) and a 0 is transmitted by a pulse – p(t).This code has the chance to go to zero before the next pulse begins, that is Polar (RZ) scheme.

- Bipolar (Return-to-Zero) : A 0 is transmitted with no pulse and a 1 is transmitted with a p(t) or – p(t) depending on whether previous 1 is encoded by – p(t) or p(t). This code has the chance to go to zero before the next pulse begins, that is Bipolar (RZ) scheme. Bipolar is also known as pseudo-ternary or alternate mark inversion (AMI).

- *On-Off* (Non return-to-Zero: **NRZ**): A 1 is transmitted by a pulse p(t) and a 0 is transmitted by a no pulse (zero signal). This code has no chance to go to zero before the next pulse begins, that is On-off (NRZ) scheme.

- *Polar* (Non return-to-Zero) : A 1 is transmitted by a pulse p(t) and a 0 is transmitted by a pulse – p(t).This code has no chance to go to zero before the next pulse begins, that is Polar (NRZ) scheme.

2.3 CELLULAR TELEPHONE (RADIO MOBILE) SYSTEM

The cellular phone service area is divided into smaller geographical areas called cells. Each cell has a base station with a tower, which receives and transmits phone signals to mobile users. The mobile telephone switching office (MTSO) connects all the base stations by telephone lines in turn is connected to the telephone central office by phone lines. A caller communicates via radio channel to a cell-site base station, which sends the signal to the MTSO. When a mobile party calls, MTSO sends the signal to the base station of the cell where the called party is located. The base station transmits the signal to the called party using the available radio channel in the cell. As the caller moves from one cell to another, the MTSO automatically switches the user to an available channel in the new cell while the call progress. The manufacturer's serial number and phone number assigned by the phone company is forwarded to the MTSO during a initialization of a call. When the call terminates, the radio channels become available for another user. The MTSO continuously monitors the signal strength of a phone call and the signal attenuation beyond some point is viewed as an indication of the caller moving from a previous cell to the next cell. MTSO also searches for a neighboring cell, where the signal strength from the caller is stronger and then automatically switches the caller to the next base station. This a rapid switching that takes the users(c) unnoticed.

2.4 CONCLUSION

Here, conclusion takes into account the treatise of digital communication concepts with the briefing of Shannon Law. It states that the information-carrying capacity of a communication link is proportional to its bandwidth. It explains the why certain technologies are used in the intelligent computer communication systems. The mobile telephone switching office is discussed as to the functional service of SMS and MMS in ICC system. The analysis of ICC system can look into detail the workings of the mobile radio system (input to processing). The design of ICC systems takes into consideration inputting and outputting. The five main functions of a modem is enumerated. The principles of data transmission is considered in this paper's concluding remarks.

2.5 FURTHER WORK

The future of the survey review will look at an intelligent communication of AT-command set for the intelligent modem. A detail implementation of the list processing of character-command computation.

REFERENCES

[1] Database System Concepts by Henry F. Korth and Abraham Silberschatz.

[2] Digital Data Communications by Jack Quinn.

[3] Modern Digital and Analog Communication

Systems, B. P. Lathi.

[4] Essential JTAPI, JAVA™ TELEPHONY API by Spencer Roberts [5] SAMS Teach Yourself UML In 24 Hours third edition by Joseph Schmuller [6] XML Bible second edition by Elliote Rusty Harold

Chapter 3
Software
<u>Organization</u>

A Compact System as a Software Organization of an Intelligent Computer Communication

FRANK APPIAH , *Softgene Technologies Institute GH. Affiliate: KWAME NKRUMAH UNIVERSITY OF SCIENCE AND TECHNOLOGY, DEPARTMENT OF COMPUTER ENGINEERING, KUMASI, ASHANTI, GHANA.*

ABSTRACT.

This software analysis paper starts with a short introduction to a compact system as an organization to intelligent computer communication in the perspective of software system. Section 3.2 studies and investigates the basic software analysis work-flow of AINRS. In there, three models are used in studying the functioning of an ICC system mainly at the entity, dynamic and functional levels of the system. In section 3.3, I conclude on the organization of an intelligent computer communication.

CCS Concepts:

Intelligent Computer Communication : software architecture ; software system : communication technology.

KEYWORDS

ACMICC2017, intelligent; computer; communication; technology; database; telephony; system; messaging; language; entity.

ACM Reference Format:

Prof. Frank Appiah. 2017. A compact system as a software organization of an intelligent

computer communication. ACM-
SIGDOC,XXXX. 12 (October, 2017), 36
pages.

3.1 INTRODUCTION

This paper will look at a compact system
mainly the structure, composition of system
and the software organization of automatic
information retrieval system (AINRS). This is
a small software system that brings the composition of computer communication system
to an unimaginable decision or perspective.
The structure of the system will begin with
the analysis of the software that makes
AINRS function to its correctness. The heart
of the organization of automatic information
retrieval system will beat in our exploration
to intelligent computer communication(ICC).
A compact system indeed but the definition
of concepts here is broad and detail enough
to replicate the software and build yourself
an ICC. With an ICC system several solution
application can be implemented.

3.2 OVERVIEW OF SOFTWARE ANALYSIS

The Java Programming language has come along a long way to be the most popular and well supported programming language available today. It provides support interfacing with hardware through the ports on the computer. It is a language that also has a lot of support for client server interactions and distribution of resources in such applications. The software analysis work-flow[7] investigates the Automatic Information Retrieval System (AINRS) , which is composed of various models each of which has specific and predefined sets of tasks. These model units are :

- Entity Class Model

- Dynamic Model

- Functional Model

This model can be regarded as the brain of this system. It houses all the primary but not specific methods that the system will need in order to understand and retrieve the necessary information the user requests for. However this unit is only an abstraction (an interface) that needs to be implemented in each case since the type of information needed by every type of application. For example in a school scenario user requesting for a CWA will have a different intelligence as compare to a user requesting for hotel reservations in a hotel scenario. As such the intelligence module (interface) will have a method such as analyse but will have different implementations for the above two scenarios.

3.2.1 Entity Class Modeling

An entity is an object that exists and is distinguishable from other objects. An entity set is a set of entities of the same type. An entity is represented by a set of attributes.

Formally, an attribute is a function which maps an entity st into a domain. These entity classes are stored in the E-R (entity-relationship) database:

- Authenticator: A class that models the database which contains the authenticator objects for allowing or denying users of the system.

- Admin: A class that models the database which contains the administrative objects for accessing all permissions.

- Cwapoint: class that models the database that contains student information relating to cwa points.

- Timetable: class that models the database that contains student information relating to time table.

- Trail: class that models the database that contains student information relating to trail courses.

- GeneralInfo: A class that models the database which contains the general-info objects.

- Student: A class that models the student in the database such details as first name, last name, index number of a student object.

- Registercourse: A class that models the student in the database with such details as the number of registered courses and their details.

A relationship is an association among several entities. On the other hand, a relationship set is a set of relationships of the same type. From the class diagram(Illustration 9), there are relations like:

- A student has many registered courses. This is a one-to-many relationship.

- A registered course has a student. This is a one-to-one relationship.

- A registered course has a time table. This is a one-to-one relationship.

- A time table has many registered courses. This is a one-to-many relationship.

- A student has many cwa points. This is a one-to-many relationship.

- A cwa point has a student . This is a one-to-one relationship.

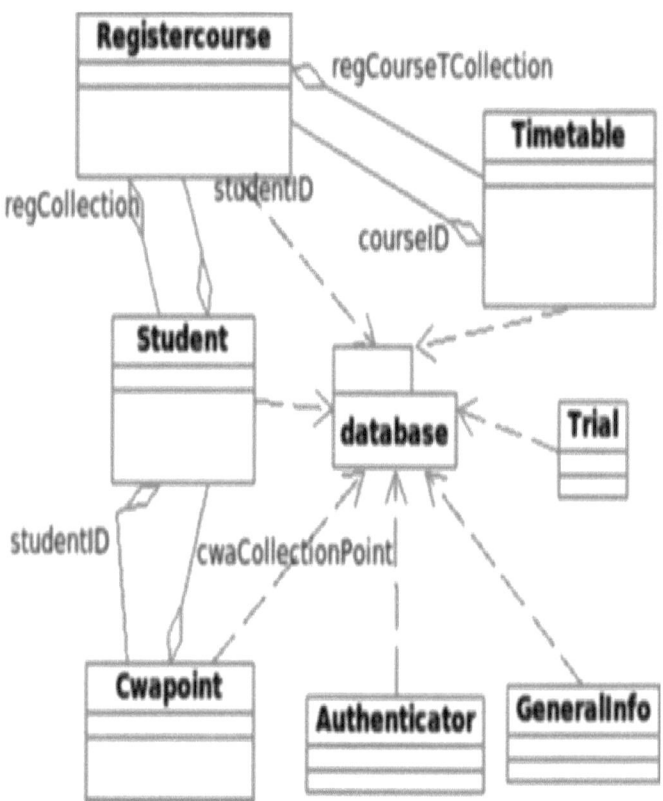

Illustration 10: Class Diagram

(c) Appiah, Frank. Class Diagrams showing associations of AINRS without attributes. 2008. Umbrello UML Modeller. Author's work [7]. Accra, Ghana. See attributes below.

E-R Diagrams

The overall logical structure of a database can be expressed graphically by an E-R diagram which consists of the following components:

- Rectangles, which represent entity sets.

- Ellipses, which represent attributes.

- Diamonds, which represent relationship sets.

- Lines, which link attributes to entity sets and entity sets to relationship sets.

ADMIN DIAGRAM

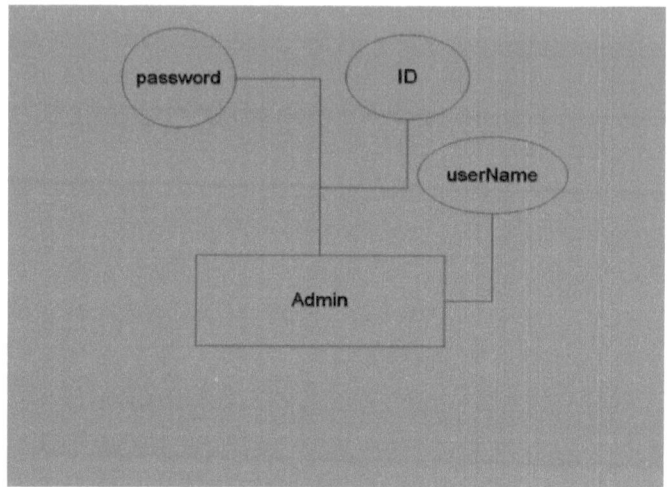

Illustration 11: Admin Diagram

AUTHENTICATOR DIAGRAM

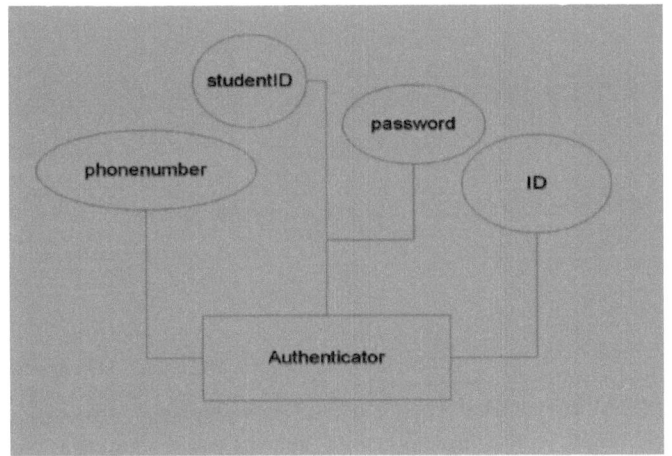

Illustration 12: Authenticator Diagram

CWAPOINT DIAGRAM

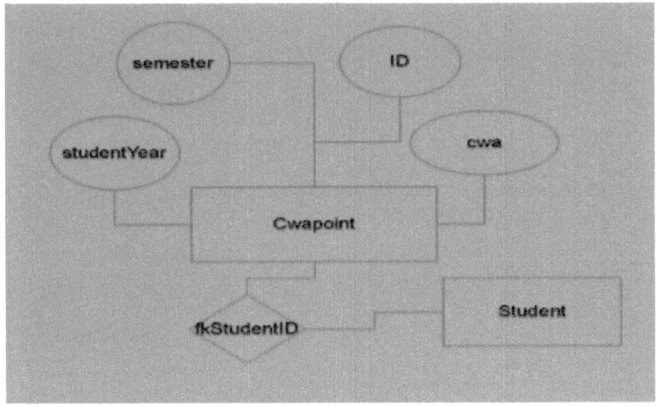

Illustration 13: Cwapoint Diagram

(c) Frank Appiah. ER Diagrams of AINRS. 2008. OpenOffice Drawing. Author's work [7]. Kumasi, Ghana.

TIMETABLE DIAGRAM

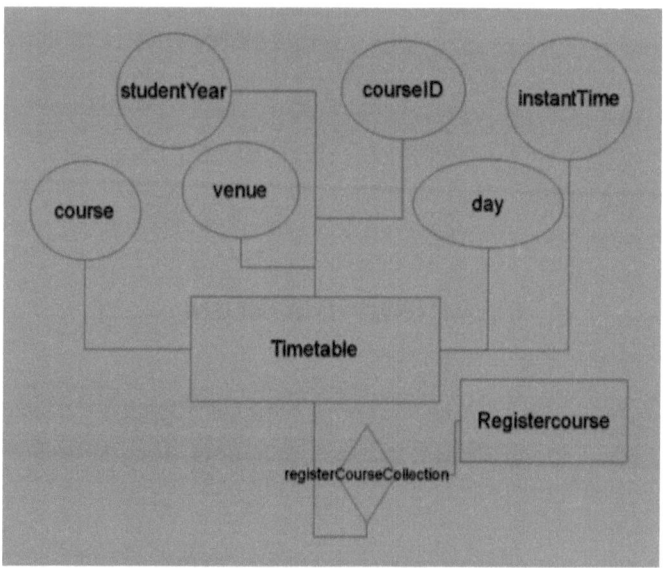

Illustration 14: Timetable Diagram

(c) Frank Appiah. ER Diagrams of AINRS. 2008. OpenOffice Drawing. Author's work [7]. Kumasi, Ghana.

TRAIL DIAGRAM

(c) Frank Appiah. ER Diagrams of AINRS. 2008. OpenOffice Drawing. Author's work [7]. Kumasi, Ghana.

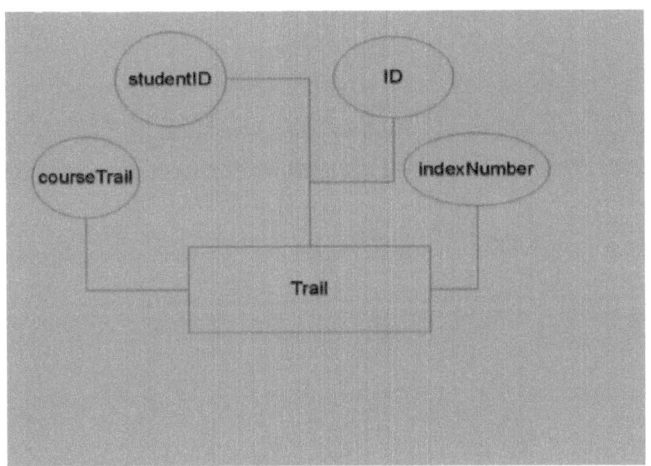

Illustration 15: Trail Diagram

GENERALINFO DIAGRAM

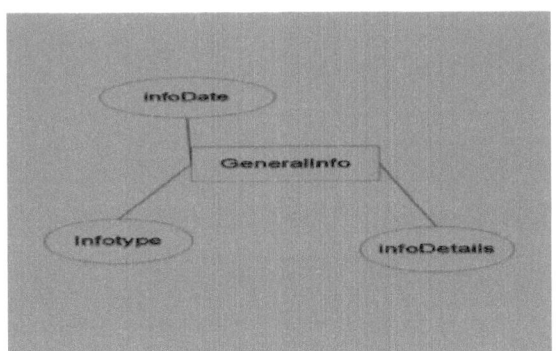

Illustration 16: GeneralInfo Diagram

(c) Frank Appiah. ER Diagrams of AINRS. 2008.

OpenOffice Drawing. Author's work [7]. Kumasi, Ghana.

STUDENT DIAGRAM

(c) Frank Appiah. ER Diagrams of AINRS. 2008.
OpenOffice Drawing. Author's work [7]. Kumasi, Ghana

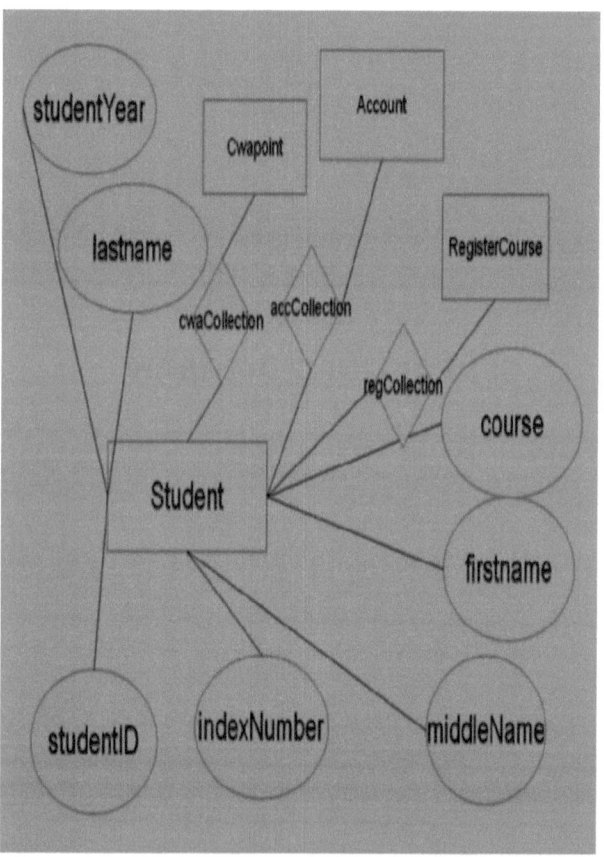

Illustration 17: Student Diagram

The remaining entities namely smssvr_in and smssvr_out can be diagrammed as the same but we will leave it here for now.

Check below for the details of the class diagrams in the Illustrations. For clarity, sometimes do check the E-R diagrams.

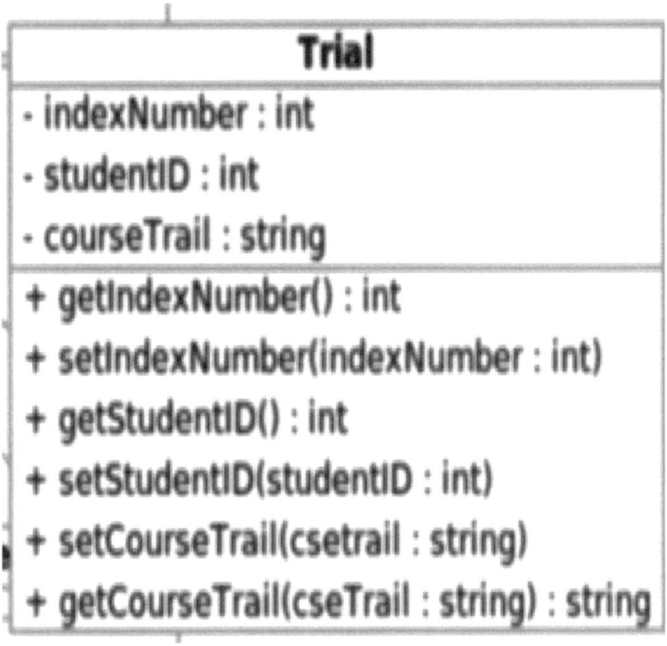

Trial
- indexNumber : int
- studentID : int
- courseTrail : string
+ getIndexNumber() : int
+ setIndexNumber(indexNumber : int)
+ getStudentID() : int
+ setStudentID(studentID : int)
+ setCourseTrail(csetrail : string)
+ getCourseTrail(cseTrail : string) : string

Illustration 18: Trail Class Diagram

(c) Frank Appiah. Trail Class Diagrams of AINRS. 2008. Umbrello UML Modeller. Author's work [7]. Kumasi, Ghana

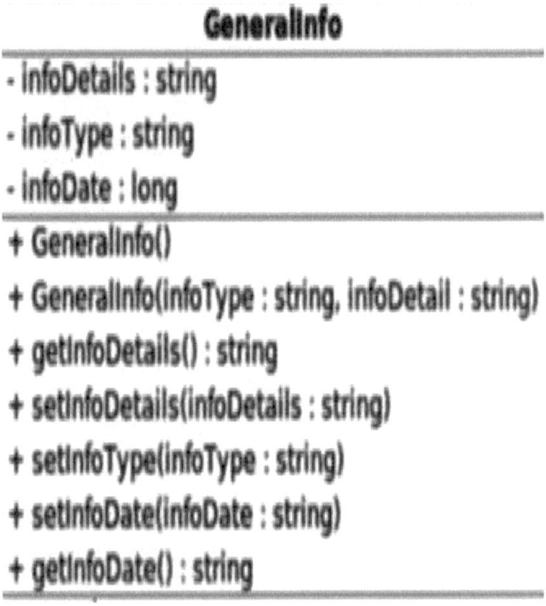

GeneralInfo

- infoDetails : string
- infoType : string
- infoDate : long

+ GeneralInfo()
+ GeneralInfo(infoType : string, infoDetail : string)
+ getInfoDetails() : string
+ setInfoDetails(infoDetails : string)
+ setInfoType(infoType : string)
+ setInfoDate(infoDate : string)
+ getInfoDate() : string

Illustration 19: GeneralInfo Class Diagram

*(c) Frank Appiah. GeneralInfo Class Diagrams of
AINRS. 2008. Umbrello UML Modeller. Author's work
[7]. Kumasi, Ghana*

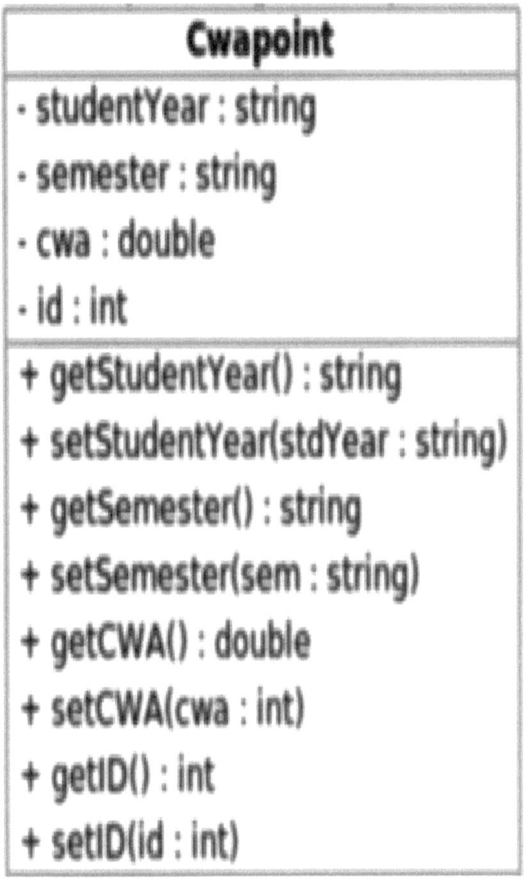

Illustration 20: Cwapoint class diagram

(c) Frank Appiah. Cwapoint Class Diagrams of
AINRS. 2008. Umbrello UML Modeller. Author's work
[7]. Kumasi, Ghana

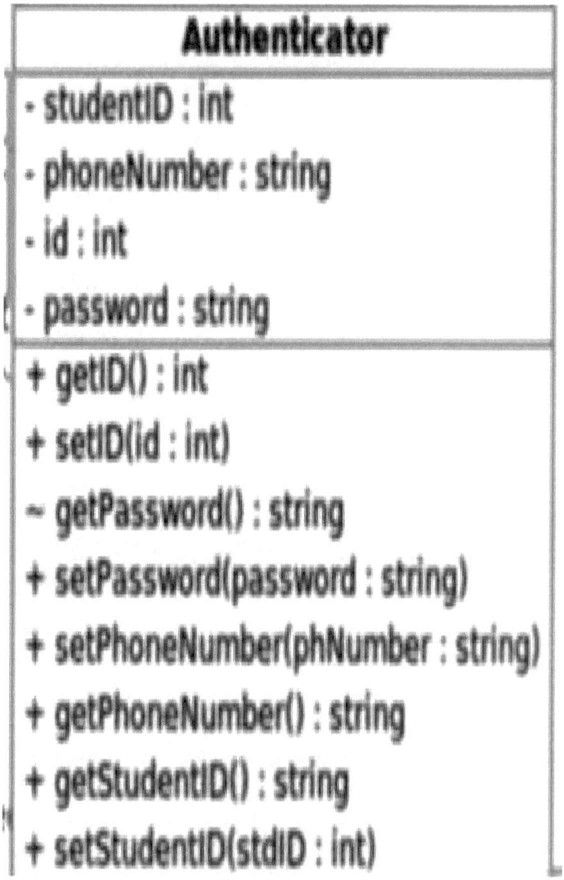

Authenticator
- studentID : int
- phoneNumber : string
- id : int
- password : string
+ getID() : int
+ setID(id : int)
~ getPassword() : string
+ setPassword(password : string)
+ setPhoneNumber(phNumber : string)
+ getPhoneNumber() : string
+ getStudentID() : string
+ setStudentID(stdID : int)

Illustration 21: Authenticator Class Diagram

(c) Frank Appiah. Authenticator Class Diagrams of AINRS. 2008. Umbrello UML Modeller. Author's work [7]. Kumasi, Ghana

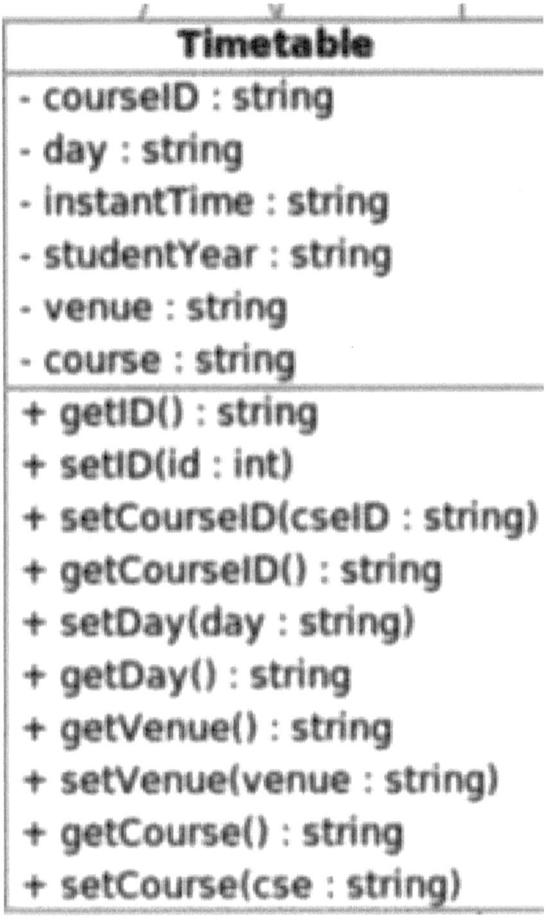

```
                 Timetable
─────────────────────────────────────────
 - courseID : string
 - day : string
 - instantTime : string
 - studentYear : string
 - venue : string
 - course : string
─────────────────────────────────────────
 + getID() : string
 + setID(id : int)
 + setCourseID(cseID : string)
 + getCourseID() : string
 + setDay(day : string)
 + getDay() : string
 + getVenue() : string
 + setVenue(venue : string)
 + getCourse() : string
 + setCourse(cse : string)
```

Illustration 22: Timetable Class Diagram

(c) Frank Appiah. Timetable Class Diagrams of AINRS. 2008.
Umbrello UML Modeller. Author's work [7]. Kumasi, Ghana

3.2.2 Dynamic Modeling OF AINRS

AINRS is basically a system to which any specific application being implemented can be attached. Illustration 17-22 is the class diagram illustrating the above relationships. The flow of communication in AINRS is very simple. When a call is made to the system either from a cellular phone or fixed line a connection is established with the telephone System. After the connection has been set-up and established, a set of instructions are played to the caller as to which combinations to press pertaining specific information by the Text-To-Speech manager. Inputs from the caller are received the DTMF recognizer in the telephone system. The inputs are then analysed by the appropriate intelligence and appropriate requests are sent to the database system from which the corresponding results are delivered to the telephone system. A structured message is constructed from the results and sent to the Text-To-Speech manager which is then played to the caller's hearing. On the other hand when a client sends in a short message, the mobile system detects that an inbound notification event has occurred. The message is opened and its contents are analysed. Invalid requests are discarded and an error is delivered. For valid requests however, the appropriate query is constructed and sent to the database System. The database system uses this query to find the appropriate information. The results are sent to the mobile system which then sends it as text to the client.

Flow of Communication into the System

The flow of communication is the dynamic modelling of the AINRS as illustrated in 24 with zoom out in Illustration 25 and Illustration 26. Just follow the sequence diagram and the functions are implemented in the enumeration below. The operations performed by or to each entity class are:

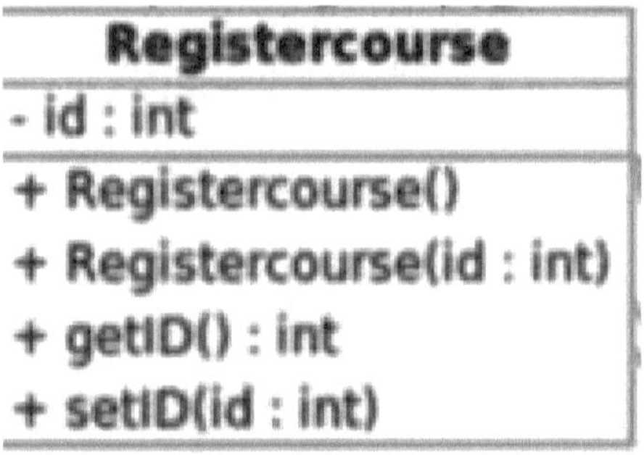

Registercourse

- id : int

+ Registercourse()
+ Registercourse(id : int)
+ getID() : int
+ setID(id : int)

Illustration 23: Registercourse Class Diagram

(c) Frank Appiah. Registercourse Class Diagrams of AINRS. 2008. Umbrello UML Modeller. Author's work [7]. Kumasi, Ghana

1. A client makes a call to the dedicated line and sends his student number. The call is established by the telephone system. *Function- **accept(call, Std-No)***.

2. The student number is then sent to the intelligence unit. With this the appro-

priate intelligence is constructed and passed to the database System for validation. *Function*- **getUserInput(call, stdNo).**

3. The database System connects to the student's database and validates the student number. *Function*- **getUserInput(stdNo).**

4. Upon validation a reply as to whether the client is valid or not is returned to the intelligence. *Function*- **processStudentNumber(stdNo)**.

5. The intelligence passes this information (once again structured) to the speech manager. When the client is valid, the instructions are read out to him otherwise he is rejected as an invalid client and the call ended. *Function*- **validUser=false or validUser=true.**

6. When the instructions are read the user is required to a make a selection either by voice or DTMF tones. *Function-* **readOut("Invalid User") on false. Or read-Out("Valid User") on true.**

7. Each selection is then forwarded to the intelligence. *Function-* **getChoice(choice)**

8. Upon analysis by the intelligence part, the client's request is structured and sent through the database System to the appropriate the database class handler. *Function-* **getChoice(choice).**

9. The reply to the request retrieved is forwarded to the intelligence part which structures it and sends the information in a properly structured way. *Function-* **retrieve(choice).**

10. Upon validation that the client has received his/her information the telephone system waits for a short interval

before terminating the call. *Function-*
Response().

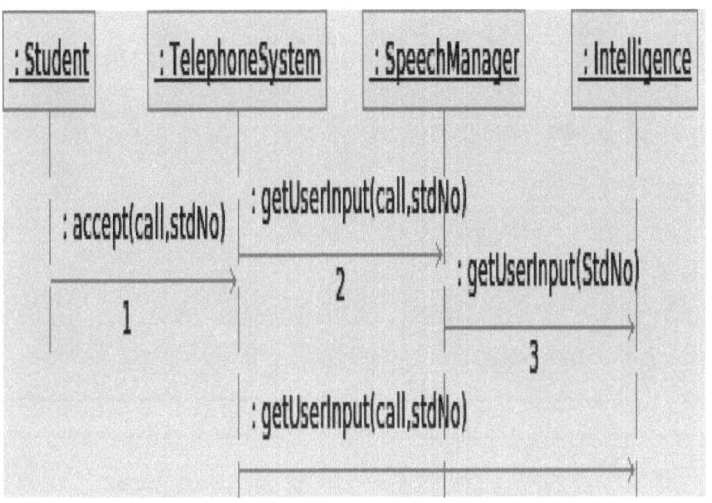

Illustration 24: 1ˢᵗ Sequence Diagram (Zoom)

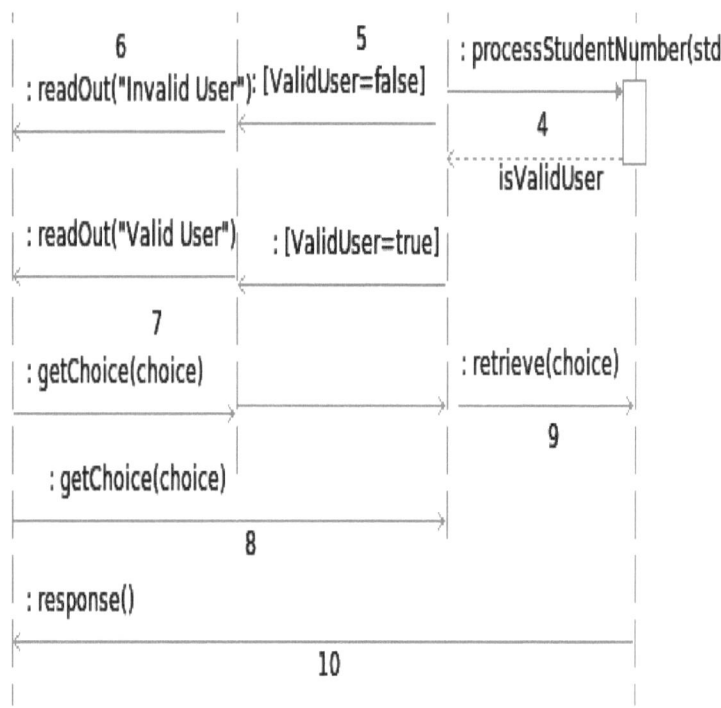

Illustration 25: 2nd Sequence Diagram [Zoom]

(c) Frank Appiah. Sequence Diagrams of AINRS. 2008. Umbrello UML Modeller. Author's work [7]. Kumasi, Ghana.

In this system application the client (student) has many choices he/she can make depending on the information he/she requires. These include

- Retrieval of CWA results

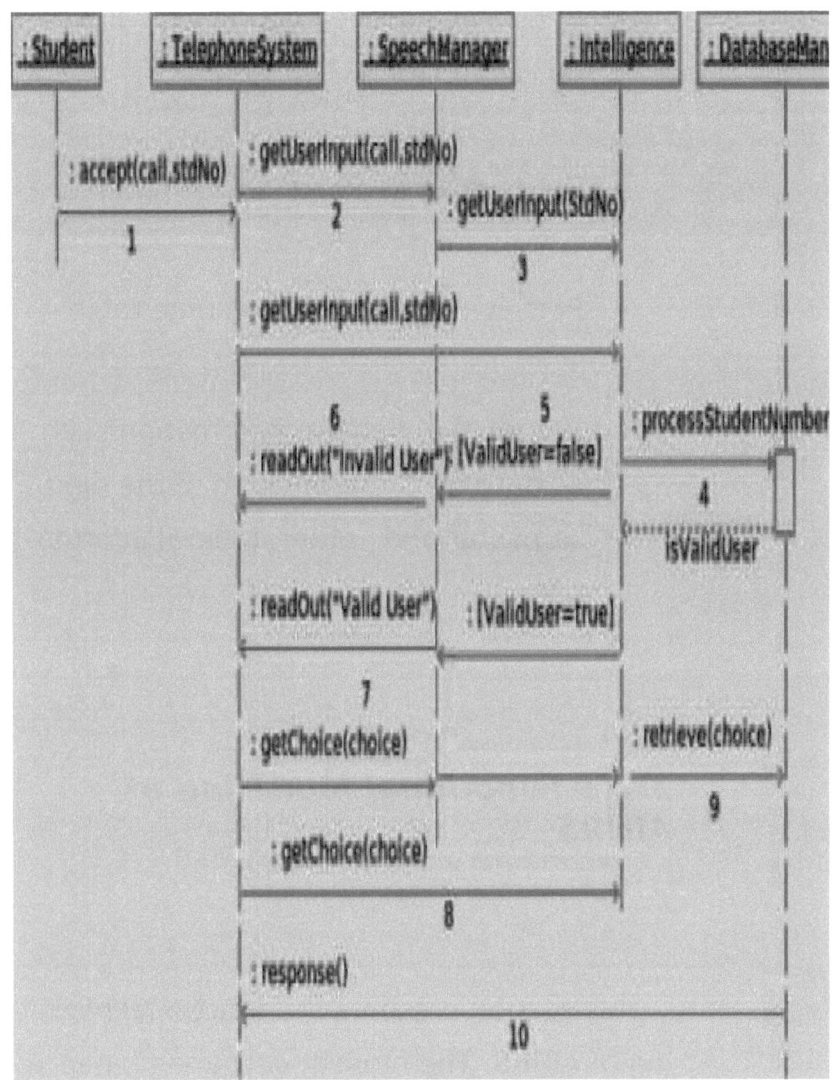

Illustration 26: Full Sequence Diagram

(c) Frank Appiah. Sequence Diagrams of AINRS. 2008. Umbrello UML Modeller.
Author's work [7]. Kumasi, Ghana.

- Retrieval of trailed courses for a previous semester

- Retrieval of semesters time table

- Retrieval of examination time table

- Retrieval of general information such as date for the reopening of the next semester and   other general information.

3.2.3 Functional Modeling of AINRS

This section presents the functional modeling of AINRS. The present scenarios (an instance of a use case) of all the use cases are in Figure 3. The figure show the various use cases for this application. An instance of a use case will be any of the following:

- Receive exams time table

- Receive trail courses

- Request trail courses

- Request general information

- Receive general information

- Request CWA

- Request course outline

- Receive course outline

- Receive time table

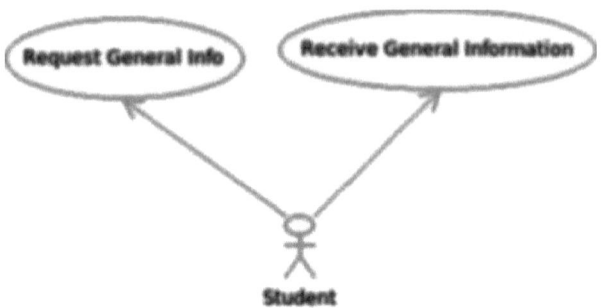

Illustration 27: Use case 1

*(c) Frank Appiah. Use case Diagram of AINRS. 2008. UML
Modeller Author's work [7]. Kumasi, Ghana.*

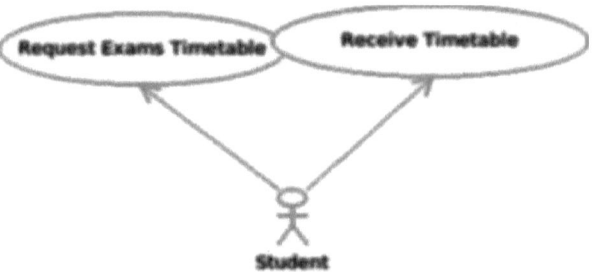

Illustration 28: Use Case 2

*(c) Frank Appiah. Use case Diagram of AINRS. 2008. UML
Modeller Author's work [7]. Kumasi, Ghana.*

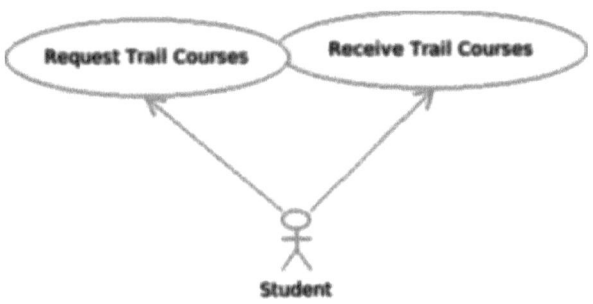

Illustration 29: Use Case 3

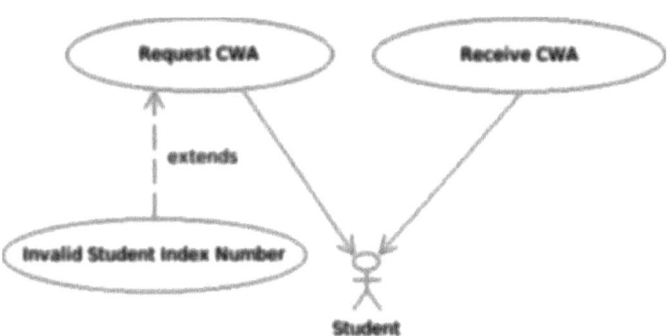

Illustration 30: Use Case 4

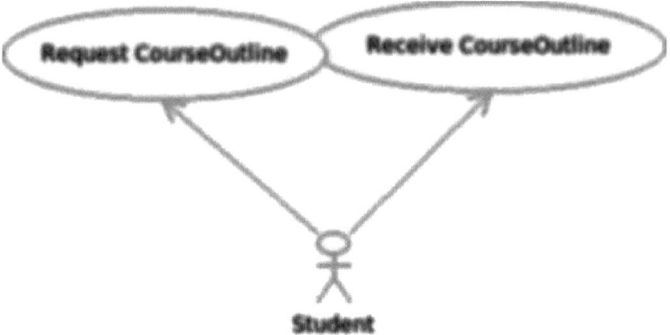

Illustration 31: Use Case 5

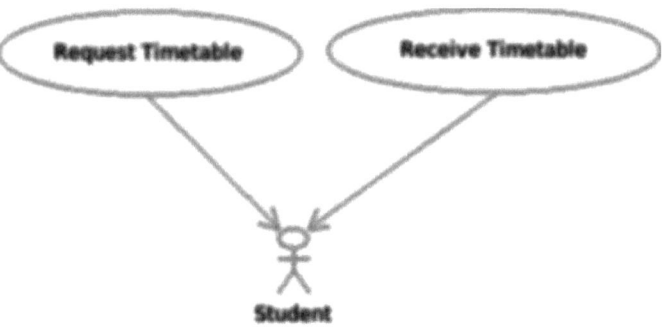

Illustration 32: Use Case 6

3.3 CONCLUSION

Here, the conclusion takes into account the treatise of software analysis concepts with the briefing on entity, functional and dynamic models. I show that taking AINRS into consideration a hypothetical ICC system can be built by analyzing a requirement specification on similar modular approach. First, the ICC system needs to decide on the entity models to be implemented in a database, second the dynamic models of the system will need to be thought through and lastly, the functional model of the system usages have to laid out. These three models can cause the heart of the system to start beating properly in terms of data communication.

REFERENCES

[1] Database System Concepts by Henry F. Korth and Abraham Silberschatz.

[2] Digital Data Communications by Jack Quinn.

[3] Modern Digital and Analog Communication Systems, B. P. Lathi.

[4] Essential JTAPI, JAVA™ TELEPHONY API by Spencer Roberts

[5] SAMS Teach Yourself UML In 24 Hours third edition by Joseph Schmuller

[6] XML Bible second edition by Elliote Rusty Harold

[7] Automatic Information Retrieval System by Frank Kwabena Appiah, 2016. Createspace Publishing Inc.

Alphabetical Index

Keyword Index

www.ingramcontent.com/pod-product-compliance
Lightning Source LLC
Chambersburg PA
CBHW030857180526
45163CB00004B/1618